Surface Crystallographic Information Service

Database and Graphing Programs
Version 1.1, January 1987

User's Manual

Surface Crystallographic Information Service

Database and Graphing Programs
Version 1.1, January 1987

User's Manual

by

J. B. Pendry
The Blackett Laboratory,
Imperial College of Science and Technology, London, U.K.

D. REIDEL PUBLISHING COMPANY
A MEMBER OF THE KLUWER ACADEMIC PUBLISHERS GROUP
DORDRECHT / BOSTON / LANCASTER / TOKYO

Library of Congress Cataloging in Publication Data

CIP

Surface crystallographic information service.
User's Manual.
J. B. Pendry.
ISBN-13: 978-90-277-2504-2 e-ISBN-13: 978-94-009-3883-0
DOI: 10.107/978-94-009-3883-0

Published by D. Reidel Publishing Company,
P.O. Box 17, 3300 AA Dordrecht, Holland.

Sold and distributed in the U.S.A. and Canada
by Kluwer Academic Publishers,
101 Philip Drive, Assinippi Park, Norwell, MA 02061, U.S.A.

In all other countries, sold and distributed
by Kluwer Academic Publishers Group,
P.O. Box 322, 3300 AH Dordrecht, Holland.

SURFACE CRYSTALLOGRAPHIC INFORMATION SERVICE
DATABASE AND GRAPHING PROGRAMS

CONTENTS

SURFACE CRYSTALLOGRAPHIC INFORMATION SERVICE
DATABASE AND GRAPHING PROGRAMS
version 1.1 January 1987

J B Pendry
Imperial College
London

1. General Description

This program is designed to provide access to the SCIS database of surface crystal structures and to present the information either in tabular form, or graphically. It can also be used to create a new database for the user and is ideal for experimenting with new surface crystal structures. Complex molecules can also be modelled and graphed.

The program is menu driven and holds surface structures on a formatted database which can be updated. There are facilities for selecting a subset of structures from the database, writing these structures to a new file, printing them out, or simply examining the structures in the context of the program

A given structure is input to the program by completing a standard structure form which can then be stored. It is possible to check the logical consistency of a surface structure with respect to overlapping or unsupported atoms. Sections of the database provide space for comments on numerous aspects of the surface structure. A database can be searched for specified substrates and/or overlayers. The graphing programs take the coordinates from the form and make a perspective picture of the structure with atomic radii obtained from an internal table and the viewing angle and distance chosen by the user, as well as the colour of the atoms. There is an option to rotate the surface about any of three axes by a selected angle. Labels can be written on the figure if a printout or photographic slide is to be made. If the structure is then filed it can be recalled at any subsequent time by reference to a "common name" assigned at the time the table was created.

The requirements for running the program are an IBM, or IBM compatible, PC or PC AT with an IBM graphics card and colour monitor; running under MS-DOS or PC-DOS 2.0 or later versions; two floppy disk drives or a hard disk.

2. Getting Started

Two disks are supplied: one containing the programs, disk 1, the other containing the database, disk 2. Before doing any work at all on the programs it is recommended that you make working copies of both disks. Put away your original disks to be used to recreate the programs in case of accidents.

2.1 EXPANDING THE DATABASE

For reasons of compactness the database is held on a compressed sequential file named SCISCOMP.DAT which must be expanded into partitioned datasets. The expanded file can be accessed much more rapidly, but is approximately five times the size of the compressed version.

If your PC has a hard disk, and assuming that your hard disk is drive C, create a separate directory for SCIS with the command

C:MD\SCIS

Then copy all the files on both disks into this directory. Make the hard disk the default drive and \SCIS the default directory. The expanded database will occupy about 1.7 Mbytes, so you should check that there is enough space by means of the DIR command in DOS. Do this after you have copied the programs and the compressed database. Type SCISDCOM to start the decompression process. Respond to the input and output file prompts with C (if this is the label for your hard disk), and type 2000 followed by the ENTER key for the capacity of the drive.

If your machine does not have a hard disk then you will need several floppy disks for the expanded database. For low density disk drives (320 Kbytes) six disks will be needed, for high density drives (1.2Mbytes) only two disks will be required. Make sure that you have the requisite number of formatted disks to hand before beginning the decompression procedure. You can check the capacity of the disks by means of the CHKDSK command in DOS. The expanded database will be held on files named $$$T1.DAT, $$$T2.DAT, etc. Label the blank formatted disks clearly and place the one labeled $$$T1.DAT in disk drive B. Then place your working copy of disk 1 in drive A, make sure that the default disk drive is drive A, and type SCISDCOM. After the disk drive has finished reading in the file remove disk 1 from drive A and replace it with disk 2. Respond to the input file prompt with A and to the output file prompt with B. For the capacity of the disk drive type the appropriate number in kilobytes, 320 for low density and 1200 for high density, followed by ENTER.

In response to the prompt for an input file type SCISCOMP.DAT followed by ENTER. Name the output files $$$T1.DAT, $$$T2.DAT, in sequence. Each one of these files will need to be initialised so type Y in response to the initialisation prompt. The program gives you several opportunities to abort if you have forgotten to complete some of the above steps. Follow the prompts, feeding formatted disks into the machine as requested by the program. All the files must be decompressed at the same time: if the process is interrupted you must return to the start once more.

If you have a hard disk, at the end of the decompression procedure you may wish to delete the files SCISDCOM.EXE and SCISCOMP.DAT from the hard disk, but retaining backup copies on floppy disks.

2.2 SETTING UP THE PROGRAM

The program is held as four EXE files on your working copy of disk 1: SCIS.EXE, $$S2.EXE, $$SG1.EXE, and $$SG2.EXE. It is in compiled basic which has been created on an IBM PC and PC AT running under PC DOS 2.0, or later versions. It requires a colour monitor and graphics card to be installed, and two floppy disk drives, or a hard disk. These files must be present on the default drive together with the files $SCRATCH.DAT and $GRATCH.DAT. If you have a hard disk then copy these six files onto the directory \SCIS created in section 2.1, if you have not already done so. The programs are now ready to use.

Before starting the program first ensure that i) you know the name and location of the data file you wish to use ii) the four EXE files plus the $SCRATCH.DAT and $GRATCH.DAT files are in the default drive. Then type SCIS in the usual way for activating EXE files.

It is recommended that you set your keyboard in the capslock mode as most of the commands require responses in capitals.

3. The Main Utilities Menu

Having set up the system as described in section 2, set your keyboard into the capslock mode and start the system by typing SCIS. The main utilities menu appears on the screen:

```
                 (C) COPYRIGHT RIEDEL 1987
                 version 1.1 January 1987
          SURFACE CRYSTALLOGRAPHIC INFORMATION SERVICE
                        UTILITIES

       CHANGE THE INPUT OR OUTPUT FILES          C

       DEFINE THE CURRENT SET OF STRUCTURES      D

       FILE THE CURRENT SET OF STRUCTURES        F

       PRINT A STRUCTURE                         P

       LIST CURRENT DATA SHEETS                  L

       EDIT, INPUT AND GRAPH                     E

       STOP                                      S

CURRENT  INPUT FILE IS:
CURRENT OUTPUT FILE IS:
```

If this menu does not appear there is a fault. You can escape from the program by typing S which returns you to DOS.

3.1 SELECTING INPUT AND OUTPUT FILES

First try typing C. a new menu appears inviting you to change the input or output files:

```
CURRENT  INPUT FILE IS
CURRENT OUTPUT FILE IS



                     NEW  INPUT FILE      I

                     NEW OUTPUT FILE      O

                     EXIT                 E
```

Try typing I. A request for the name of the new input file appears. Place the disk containing your decompressed database, $$$T1.DAT, into a convenient drive. Then type the drive name followed by the files name. For example if the database is in drive B: type B:$$$T1.DAT followed by pressing the ENTER key. You can also use directories to identify a file if your DOS version allows this. Next try pressing O. At the prompt for the new output file, type $RUB.DAT followed by pressing the ENTER key. Next there is a prompt for initialisation of the output file. It is also possible to declare the same file for input and output. Since this is a new file type Y. The file will then be opened for output and declared to contain no structures. Initialising an existing file prevents you from having further access to the structures it contains - they are effectively lost. If you wish to add new surface structures to an existing file, type N. After initialisation type E to return to the main utilities menu.

When SCIS is first entered the default settings for the file names are the ones assigned when the program was last used.

3.2 DEFINING THE CURRENT SET OF STRUCTURES

At the main utilities menu type D which results in the following sub-menu:

THERE ARE 0 CURRENT STRUCTURES

LIST THE CURRENT STRUCTURES L

SELECT NEW CURRENT STRUCTURES N

RETURN TO THE UTILITIES MENU R

To view all the structures on the current input file press N. You are asked whether you wish to add the new files to the existing set: type N. Next two requests appear in sequence: the first for selection of a substrate material, the second for an adsorbate material. Press ENTER in response to each request. This tells the program to admit all substrates and all adsorbates. The structures are presented one screenfull at a time followed a by prompt:

TYPE A: ACCEPT ALL; R: REJECT ALL; S: SELECT SOME

Press R after each screen until the sub- menu reappears.

Now we shall select some structures. Press N followed by N again at the second prompt. At the request for a substrate material type Al followed by pressing the ENTER key, then at the prompt for an adsorbate press Na followed by the ENTER key. The elements must be entered in exactly this form of capital first letter followed by small second letter, with no spaces. In the case of a compound, type the accepted chemical formula, eg for sodium oxide type Na2O. Pressing the ENTER key without first typing in a formula makes the program assume that you want to see all available substrates, or all available adsorbates.

After entering the adsorbate the selected files are shown one screen full at a time. They are referred to by a common name which roughly identifies the structure. At the end of each screen full of structures the program pauses and asks whether you want to accept, reject or select some of the structures. In this case press A after the first screen of structures. After the last screen you are returned to the sub- menu.

Suppose you also wish to study the silicon surfaces. At the sub-menu press N to select some more files and at the next prompt press Y to add the silicon structures to the Aluminium structures already selected. Type Si then press ENTER for the substrate and press ENTER once more to admit any adsorbate structure. The new files are shown one screen at a time. At the first screen press S to select individual files. A large blinking cursor appears against the first structure. The cursor can be moved from structure to structure with the cursor control keys: "arrow up" and "arrow down". Select the Si(111) (1X1) LASER ANNEALED structure using the "home" key to mark. Other structures can be marked and subsequently unmarked by using the "end" key. When you have finished processing each page press the "PgDn" key to proceed to the next page.

If many structures are included in the list, there may be a pause at the end of the selection procedure as the program completes some internal housekeeping.

On returning to the sub- menu press L to check that the correct structures have been selected. Some experimentation with the structure selection process should quickly bring familiarity.

If you decide that there are too many aluminium structures, press N at the sub- menu followed by Y to retain the existing structures. Then select the aluminium substrate by Al followed by the ENTER key and the sodium overlayer by Na followed by the ENTER key. All the structures in this category are listed one screen full at a time. The silicon structures remain part of the selected structures but are not displayed at this time. At the first screen type S. The blinking cursor appears positioned at the first structure, as before, but now the previously selected structures are shown with asterisks already placed against them indicating that they are selected structures. You can now use the "home" and "end" keys to mark and unmark. Mark the first structure and unmark all the rest.

Repeat the exercise for silicon, selecting only the Si(111) (1X1) LASER ANNEALED structure.

After checking with the L command that only two structures have been selected, at the sub- menu press U to return to the main utilities menu.

3.3 FILING THE CURRENT SET OF STRUCTURES

At the main utilities menu type F to enter the filing sub- menu. If you have declared $RUB.DAT to be an output file in section 3.1 and selected two structures in 3.2, the menu will appear as follows:

```
          CURRENT FILE IS   $RUB.DAT

    NUMBER OF STRUCTURES FOR FILING IS  2


            CONTINUE FILING       C

            EXIT                  E
```

If no output file has been declared, pressing C will result in the message "NO OUTPUT FILE" appearing in the top left hand corner of the screen. Exit and enter the file selection routine.

Press C to file the two structures. The program returns you to the main utilities menu after filing.

3.4 PRINTING A STRUCTURE

At the main utilities menu type P. If you have selected two structures in 3.2, the sub- menu will appear as follows:

CURRENT FILE IS $$$T1.DAT

NUMBER OF STRUCTURES FOR PRINTING IS 2

CONTINUE PRINTING C

EXIT E

Make sure that your printer is switched on and connected to the PC. Pressing C causes the printer to move to the top of a new page, then to print the two structures, move to the top of the next new page and finally return control to the main utilities menu. Pressing E brings back the main utilities menu without any printing. If a printer is not connected an error exit from the program will result. Restart the program by typing SCIS.

3.5 LISTING A STRUCTURE

At the main utilities menu type L. If you have selected two structures in 3.2, the sub- menu will appear as follows:

CURRENT FILE IS $$$T1.DAT

NUMBER OF STRUCTURES FOR PRINTING IS 2

CONTINUE LISTING C

EXIT E

Pressing C causes the first page of the first selected structure to appear. To move to the next page, press the ENTER key. After all the pages have been displayed for the first structure the following prompt appears:

ACCEPT OR REJECT THE CURRENT FILE - A OR R?
TO EXIT - TYPE E

This gives you a second chance to adjust your selection of files based on the more complete information which has just been listed. Pressing E brings back the main utilities menu; Pressing A or R accepts or rejects the structure just displayed and moves on to the next structure.

There is also a facility for viewing and selecting structures in the editing, input , and graphing routines.

4. The Main Editing and Input Menu

Having selected "E" at the main utilities menu the program pauses for a moment and then presents the main editing and input menu. If you have selected $$$T1.DAT as the input file and $RUB.DAT as the output file the screen shows:

```
              SURFACE CRYSTALLOGRAPHIC INFORMATION SERVICE

                 EDITING, INPUT AND GRAPHING FACILITIES


                  INPUT A NEW STRUCTURE                 N

                  REVISE AN OLD STRUCTURE               O

                  RETURN TO UTILITIES                   R




CURRENT  INPUT FILE IS:    $$$T1.DAT
CURRENT OUTPUT FILE IS:    $RUB.DAT
```

The input facility is moderately complex and will be described separately in section 6.

4.1 SELECTING AN OLD STRUCTURE

The structures picked from the input file in the utilities section are carried forward to this section. It is possible to select any one of them for detailed revision or for graphing.

Pressing O produces a list of the current structures. If you have followed the selection procedure given in section 3.2 the Al(100) C(2X2) Na and Si(100) (2X1) structures will be listed in that order. Type 1 then ENTER to select Al(100) C(2X2) Na and the first page of the standard structure form for Al(100) C(2x2) Na appears. Although it is not evident at this stage there are two types of entry in the form: box labels which are the same for every form and cannot be changed, and box entries which have to be entered for each structure. The important parts of the form are the COMMON NAME box, which serves as the label by which you will subsequently identify the structure and therefore must be carefully chosen; the REFERENCE UNIT CELL, which specifies the lengths of the surface cell sides and the angle between them; and the table of layers. The atoms are identified by the layer in which they appear (A1, A2... for adsorbates, S1, S2... for substrate atoms) and the atom positions are specified in units of the surface cell sides. The normal layer spacing is the distance to the next layer inside the crystal given in this instance in Ångströms. The rest of the table is self explanatory except for the entry labeled R( ) which refers to the R-factor of the structure determination and is a measure of its quality.

Pressing ENTER produces the next page on the screen. A second ENTER gives the third page and a third ENTER results in the subsidiary revision menu.

```
**************************************************
* REMEMBER TO FILE ANY CHANGES IN THE STRUCTURE *
**************************************************

          REVISE STRUCTURE               R
          FILE                           F
          TEST FOR INCONSISTENCIES       T
          ENTER GRAPHS(NO BONDS)         G
          ENTER GRAPHS(WITH BONDS)       B
          NEXT STRUCTURE                 N
```

4.2 REVISING A STRUCTURE (STARTING FROM THE SUBSIDIARY MENU)

Press R and the program will show the first box in the table: the COMMON NAME box. The cursor is located at the beginning of the box. To proceed through the table press ENTER, or to go back press ESC. Pressing any key other than ENTER or ESC deletes all the contents of the current box. Step backwards and forwards through the table to get the hang of it, in particular the progression from page to page, but be careful at the very end of the table: the last ENTER will return you to the subsidiary menu. You can easily get back to the table by pressing R again. Try doctoring a few of the entries. The program will not let you type outside the box. Provided you do not file the results of your experiment you will not mess up the stored record of the structure.

When you are through experimenting use ENTER to step through to the end of the table and to the subsidiary menu. Pressing N returns you to the main editing and input menu.

4.3 FILING

From the main editing and input menu follow the instructions in 4.1 to select the Al(100) C(2X2) Na structure. At the subsidiary revision menu shown at the end of 4.1 press F to give the following menu:

```
          ADD STRUCTURE TO OUTPUT FILE          A

          REPLACE STRUCTURE IN INPUT FILE       R

          EXIT                                  E

CURRENT  INPUT FILE IS:    $$$T1.DAT
CURRENT OUTPUT FILE IS:    $RUB.DAT
```

The selected structure may be filed either as a new structure in the output file (press A) or, if only a revision of an existing structure is intended, in the input file (press R). After filing the structure the program assumes that you have finished the revision process and returns to the main editing and revision menu. Do not press R unless you are sure that you wish to change the master copy of the structure: you will not be able to retrieve the original.

Note that in contrast to the filing process shown in 3.3, only the one selected structure is filed.

Pressing E returns the program to the subsidiary revision menu without filing.

4.4 TESTING FOR INCONSISTENCIES

Follow the instructions in section 4.1 to get to the subsidiary revision menu shown at the end of 4.1, with Al(100) C(2X2) Na as the selected structure. Try checking the structure by typing T which produces a listing of coordinates and atomic radii extracted from the table. The computer then checks that no atom overlaps with its neighbours either in the same surface cell or in the cell next door, to within 10% of the Pauling radii, a table of which is held in the program. It also checks that every atom is in chemical contact with at least one other atom to within 10%. The program should return:

NO ERRORS FOUND

Press any key and the subsidiary menu reappears.

We now want to do a controlled experiment on the structure by introducing some deliberate mistakes for the test sequence to find. Return to the revise mode by pressing R at the subsidiary revision menu.

Next press N to pick another structure and follow the instructions in 4.1 to select Al(100) C(2X2) Na. Check this structure by typing T and a list of errors will appear. Sodium has some ionic character on the surface of aluminium and the Al-Na separation is slightly shorter than the metallic radii imply. Press any key and the subsidiary menu returns

Use ENTER to step through the table: move to the NORMAL LAYER SPACING column in row S3 and change the entry from 0.0 to 3.0. This pushes layer S4 out of chemical contact with the rest of the surface. Step through the rest of the form using ENTER and at the subsidiary menu type T to enter the tests. Note how the errors are logged by the program.

The internal table of atomic radii is only an approximate guide to actual bond lengths and sometimes correct structures will show errors in the testing routine.

Next we look at the graphing part of the program. Type any letter to return to the subsidiary revision menu then N to return to the main menu. Press G to select the graphing routine without bonds.

5. Using the Graphing Program

If you have just completed the steps at the end of section 4.4 then obtain an unedited copy of the standard form: starting from the main editing and input menu, press O followed by 1 at the prompt, then

ENTER. Press ENTER to work through the pages to the subsidiary revision menu shown below.

```
***************************************************
* REMEMBER TO FILE ANY CHANGES IN THE STRUCTURE *
***************************************************

          REVISE STRUCTURE            R
          FILE                        F
          TEST FOR INCONSISTENCIES    T
          ENTER GRAPHS(NO BONDS)      G
          ENTER GRAPHS(WITH BONDS)    B
          NEXT STRUCTURE              N
```

There are two graphing routines which are identical except that one draws smaller atoms connected by bonds. Generally speaking, covalent materials are best displayed with bonds, whereas metallic and ionic materials are too highly coordinated for the bonds to be drawn in clearly. The routine with bonds is slower and will accommodate less atoms than the other routine.

5.1 FIRST ENTRY INTO THE GRAPHING ROUTINES

After selection of a structure and entry into the the graphing routines from the subsidiary menu the atomic coordinates are reproduced on the screen followed by the query

DO YOU WISH TO SHOW THE 8 NEIGHBOURING CELLS - Y OR N

The options are either to view a single unit cell of the surface, or to view a unit cell plus its eight neighbours. The latter option gives a much better feel for the surface structure and its symmetry, but the former option can be of use if the surface cell is too large for nine cells to be displayed (which can happen in the bond- display mode), if you desire more rapid execution of the graphing commands, or if you are using the programs to display a molecule without reference to any surface structure. Press Y and the system prompts you for the distance from the viewpoint.

The system has selected one of the atoms close to the centre of the cluster to be the atom displayed at the centre of the screen. The centre of this atom is the point about which we shall rotate the cluster. You must now select the distance of this central atom from your viewing point. This selection controls the perspective and essentially sets the scale of the distance from your eyes to the screen. The units are the same as those used for the surface cell sides on the standard structure form. Type 30 ENTER which will give a good perspective on this structure.

In response to the prompt to change the colours press N.

The PC now has a lot of work to do. It must rotate the cluster of atoms to obtain the new coordinates, and then sort the atoms in order of their distance from your eye in order that hidden atom removal can be

performed. Hence there is a short pause followed by the prompt
MAGNIFY BY FACTOR=?
The number you type determines the size of the cluster on the screen. Type 1 then ENTER and the cluster exactly fills the screen. Pressing any key removes the cluster from the screen and returns you to the main graphs menu:

EULER ANGLES	E
CHANGE CENTRAL ATOM	V
DISTANCE	D
MAGNIFICATION	M
REMOVE ATOMS	X
COLOUR	C
PERSPECTIVE	P
LABEL FIGURE	L
ROTATE FIGURE	R
RETURN TO TABLES	T

5.2 CHANGING THE EULER ANGLES

There are two ways in which the cluster can be rotated: either by use of the keyboard to rotate about the X, Y, Z, axes by declared amounts, or by defining a set of Euler angles with respect to the original orientation of the cluster. In practice you will find the former method much the most convenient as the Euler angles are obscure and difficult to work with. Press E and the PC responds with a request for the three Euler angles to be given in degrees

PH rotates about the Z-axis to axes X' Y' Z'
TH rotates about the Y'-axis to axes X'' Y'' Z''
PS rotates about the Z''-axis

Respond to the three prompts with:
45 ENTER 54.7356 ENTER 0 ENTER
You will then be asked for a magnification factor. Respond with
1 ENTER
This choice of Euler angles rotates the Al(001) surface from the original configuration having the 4-fold axis of the bulk normal to the screen, to having the (111) axis normal to the screen. The close packing of the substrate should now be evident

5.3 CHANGING THE CENTRAL ATOM

The centre of the screen is occupied by an atom which in the first place is chosen to be as close to the centroid of the cluster as possible. All rotations take place with respect to an axis passing through the centre of this atom. To change the central atom press any key to return to the main graphs menu, then press V. You should then receive a message that the default central atom is number 3 followed by a list of the atom numbers; their X, Y, Z coordinates; their radii; and the chemical symbol of the element in question. The system invites you to choose the number of the atom to lie at the centre of the screen, or to print more atomic coordinates. Type ENTER twice to see the remainder of the list, then type 41 ENTER. Set the distance to 30 and the magnification to 1. The new plot is shifted to the right of the screen. Press any key to return to the main graphs menu then press V. After the first screen of atoms enter 3 to reselect the original central atom. Enter 30 for the distance and 1 for the magnification.

5.4 CHANGING THE DISTANCE

The distance of the viewer measured from the central atom defines the perspective, that is to say the relative sizes of the atoms at different distances from the viewer. Of course you have not yet told the PC where you are located relative to the screen so some choices of distance may produce rather curious looking objects because your viewpoint does not coincide with the one implied by the setting of distance. This can be fixed up by the perspective command reviewed in 5.8.

Press any key to return to the main graphs menu, then press D. You are prompted for the distance to be given in the same units in which the cell sides are entered into the standard table. I suggest 15, 10, 5. Terminate this experiment with a setting of 30 for the distance in preparation for the other parts of this section.

5.5 MAGNIFICATION

Setting the magnification equal to 1 just allows the cluster to fit onto the screen. Other values are possible: try pressing any key to return to the main graphs menu followed by M to obtain the magnification prompt. Experiment with various values: I suggest 0.3, 1.5, 4, 10 as a start. This option is useful for scrutinising fine details of the structure.

Finally set the magnification to 1 before leaving this section.

5.6 REMOVING THE ATOMS

Because the atoms in the cluster are drawn as solid objects, internal details of the cluster are sometimes hidden from view. In this case it is possible to remove atoms in order to see what is underneath them.

From the main graphs menu press X. The system will tell you how many atoms are currently removed from the cluster and invites you to declare a new number to be removed. The program always removes atoms starting with the closest to the viewpoint. Try removing 4, 9, 21, atoms thus revealing several of the close packed (111) planes of the nickel substrate.

Reset the number of removed atoms to 0 before leaving this section.

5.7 COLOUR

The information conveyed by a picture is much enhanced by being able to pick out different atoms in colour, set off against an appropriate background. To enter the colour selection routines from the main graphs menu press C.

A list of possible background colours appears with their identifying numbers, followed by a statement of the current background colour. The program operates in the medium resolution mode of an IBM PC with a choice of 8 background colours, and a rather more limited choice of foreground colours from two palettes each of three colours. Type the number of your choice and press ENTER.

Next there is a choice of palettes. Type the number of your choice and press ENTER.

Finally the selection of atom colours is made. Here some information about how the program draws the pictures is relevant. The atoms are drawn in order of their closeness to the viewpoint by first painting a circle in the chosen colour. Having completed the painting the program draws two delimiting circles: an outer one in the background colour, then a second circle inside the first in colour 3 from the current palette. Choice of colour 3 for an atom means that its delimiting circle blends with the paint making the atom less distinct, and less "real" in appearance on the screen. The best choice for the colour of an atom is 2, followed by 1, then 0, the background colour.

Follow the prompts to set the atom colours selecting different colours for Al and Na. Use the END key to terminate the colour selection procedure.

Press 1 ENTER at the prompt for the magnification factor, and the picture will be redrawn in the new colours. Experiment with the various options and note their advantages and limitations.

5.8 PERSPECTIVE

Once you have chosen the distance of the cluster and its magnification there is only one distance for your eye from the screen which gives a correct perspective. In practice the eye is quite tolerant of errors in perspective and a figure correct only for a given viewing distance can still make visual sense seen from other distances from the screen or, equivalently, for different magnifications. It is in any case useful during setting up of the picture to be able to fit the whole cluster on the screen.

By returning to the main graphs menu and pressing P, the program will automatically fix the magnification so that the perspective is correct when the screen is viewed from a distance of 40cms. To obtain correct perspective and have the cluster the desired size on the screen it is necessary to experiment a little with the distance setting (see 4.4), remembering to reset the perspective each time the distance is changed. A good choice of distance for most surfaces is 30Å. Very large values of the distance will result in the structure being too small to be seen on the screen!

5.9 LABELING THE FIGURE

Sometimes if the figure is to be printed or photographed for display purposes it is useful to be able to add labels. You can do this by pressing L at the main graphs menu.

The label's position is defined by the position of the first letter of the label. Since the program works in medium resolution graphics the possible rows are 1 to 25, and the columns 1 to 40. For example, press A to add a label, then type

10,1 ENTER

to define the position then at the next prompt

SODIUM ON ALUMINIUM

followed by ENTER. The label is displayed in the list at the top of the screen. Press A again, followed by

11,1 ENTER

to locate the next label, then

(100) SURFACE

followed by ENTER.

To see how these labels fit onto the picture press F and the figure will be redrawn including the label. Some experimenting may be necessary to obtain the optimum label location by re - entering the label routine and using the move command.

Before going on to the next step delete all the labels and return to the main graphs menu.

5.10 FREE ROTATION OF THE FIGURE

This is perhaps the most useful option in the menu, because the facility of rotation enables the full three dimensionality of the cluster to be explored. From the main graphs menu press R to set the parameters.

The first prompt invites you to set the number of degrees by which each command will rotate the cluster. Next some information appears about which keys control rotation about which axis. The Z axis is assumed normal to the screen, pointing outwards, the X axis points to the right of the screen, and the Y axis points upwards, completing a right - handed set. Press any of the control keys to initiate the free rotation mode. The program remains in the free rotation mode until any key other than the specified control keys is touched. Thereafter the orientation of the surface is preserved until the central atom is changed or the distance altered, in which case the original orientation showing a plan view of the surface is restored.

Press any key to exit and return to the main menu.

6. Completing the Standard Structure Form

6.1 GENERAL DESCRIPTION OF THE FORM

The data for a given surface structure are held in a standard structure form which must be completed before the structure can be viewed. The form is designed to act as a source of general information on surface structures, the techniques used for their determination and their reliability. For some purposes parts of the form will therefore be redundant and there is no need to complete them, but other parts are essential to the working of the graphing routines.

Figure 1 shows the standard structure form before completion with boxes drawn around the cells into which the user can type text or numbers. Figure 2 shows the standard form completed for Ni(100)-C(2X2)O to give a general idea of how to fill in a form.

To enter a new structure first make sure that the relevant data files are on the default directory (see section 2) then start the program by typing SCIS. The main utilities menu appears: follow 3.1 to select the desired input and output files. The same file can be declared for input and for output. Next press E to access the main editing and input menu, followed by N to begin the input of a new structure.

6.2 COMPLETION OF THE OPTIONAL BOXES

Only some of the boxes are essential to the operation of the graphing routines: namely the set beginning with,

REFERENCE UNIT CELL a= ☐

and ending with the last set of six entries under the headings
LAYER ATOM ATOM POSITIONS NORMAL LAYER SPACING ERROR
(six because there are two entries under atom positions). We consider these boxes in the next section.

The rest of the boxes can be completed in any fashion desired. However it is always advisable to complete the first entry by giving a COMMON NAME to your structure because it is by the contents of this box that the structure is identified. For economy of space the STRUCTURES EXAMINED box will terminate if you begin a line with the ENTER key, but can be no more than 5 lines long; the COMMENTS box will terminate if a line begins with either a space or an ENTER, but can be no more that 12 lines long.

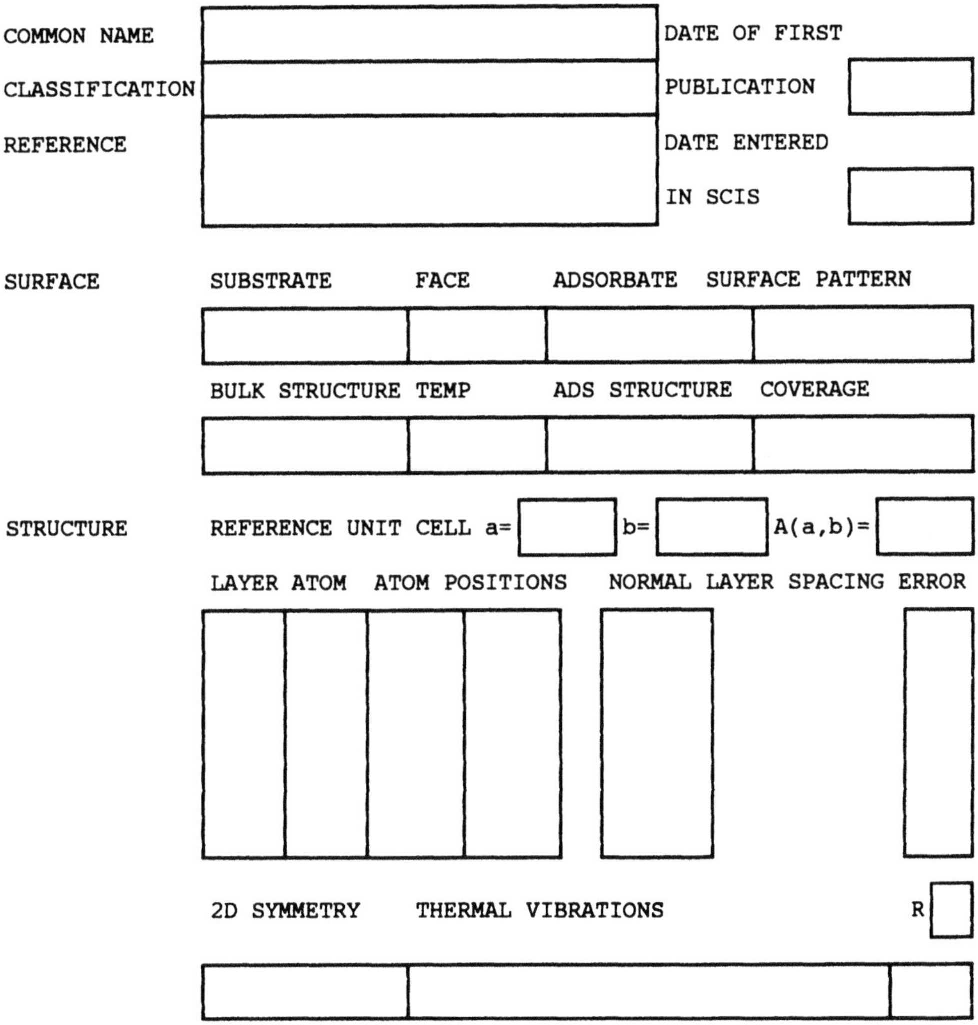

Figure 1a The standard structure form - first page - showing the spaces available for entries as boxes.

SURFACE PREPARATION	
CLEANLINESS	
METHOD OF STRUCTURAL DETERMINATION	
EXPERIMENTAL DATA	
STRUCTURES EXAMINED	

Figure 1b The standard structure form - second page.

COMMENTS

Figure 1c The standard structure form - third page

COMMON NAME	Ni(100) C(2X2)O			DATE OF FIRST	
CLASSIFICATION	28.8.48			PUBLICATION	28 NOV 1983
REFERENCE	TS RAHMAN DL MILLS JE BLACK JM SZEFTEL				
	S LEHWALD AND H IBACH,			DATE ENTERED	
	PHYS REV VOL B30 589 (1984)			IN SCIS	JULY 1985
SURFACE	SUBSTRATE	FACE	ADSORBATE	SURFACE PATTERN	
	Ni	(100)	O	(1,1/-1,1) C(2X2)	
	BULK STRUCTURE	TEMP	ADS STATE	COVERAGE	
	FCC		ATOMIC	-	
STRUCTURE	REFERENCE UNIT CELL a=3.52		b=3.52	A(a,b)=90	

LAYER	ATOM	ATOM POSITIONS		NORMAL LAYER SPACING	ERROR
A1	O	0.5	0.0	0.9	
S1	Ni	0.0	0.0	0.0	
S2	Ni	0.5	0.5	1.7	
S3	Ni	0.5	0.0	0.0	
S4	Ni	0.0	0.5	1.76	

2D SYMMETRY	THERMAL VIBRATIONS	R(-)
P4M	-	-

SURFACE PREPARATION: SURFACE CLEANED BY NEON ION BOMBARDMENT, FOLLOWED BY ANNEALING AND REPEATED SPUTTERING AND ANNEALING

CLEANLINESS: AUGER AND EELS SPECTROSCOPY

METHOD OF STRUCTURAL DETERMINATION: CALCULATION OF SURFACE PHONON SPECTRA INCLUDING ADSORBATE LAYER AND FIRST THREE LAYERS OF Ni ATOMS, ASSUMING SHORT RANGE CENTRAL FORCE AND DIPOLE-DIPOLE INTERACTIONS

EXPERIMENTAL: OFF- SPECULAR EELS DATA

STRUCTURES EXAMINED: O - Ni LAYER SEPARATIONS OF 0.9 AND 0.26 ANGSTROMS IN 4 FOLD HOLLOW SITE; PSEUDO-BRIDGE ADSORPTION SITE (0.3 ANGSTROM DISPLACEMENT ALONG (100) DIRECTION)

COMMENTS: AUTHORS CONCLUDE THAT C(2X2) AND P(2X2) ADSORBATE-SUBSTRATE RELATIONSHIP IS THE SAME, THOUGH O-Ni FORCE CONSTANT IS GREATER IN THE C(2X2) STRUCTURE BY A FACTOR OF 0.55

Figure 2 The standard structure form shown completed for Ni(100) C(2X2)O. The form has been compressed into one page.

Data are entered simply by typing text into the box. The backspace key deletes the previous letter and positions the cursor one space back. On reaching the end of the space allowed for that box the PC will beep and refuse to accept any further characters. Pressing ENTER records the entry and moves the cursor to the beginning of the next box.

If you wish to change a box that has already been ENTERed,it is possible to step back through the boxes one at a time by pressing the ESC key. You can hop forwards and backwards through the form by use of the ENTER and ESC keys but pressing any other key deletes the entire contents of a box once it has been ENTERed, and resets the cursor to the beginning of the box. The space available for entries can be judged from figure 1.

The standard form is divided into between three and five pages depending on how many coordinates are entered (see the next section) and the program is informed that the coordinate list has been completed by entry of a blank line of coordinates. Only as many pages are presented as are needed to display the coordinates. At the end of each page, as defined by figure 1, the cursor pauses at the bottom of the page and an extra ENTER is required to move to the next page.

Moving forward through the form for the first time, the boxes and their labels are revealed one at a time, and stay on the screen when the ESC key is used to step backwards. This is even true when stepping back to a previous page, except in the case where there are so many coordinates that the form contains five pages. Then the screen memory does not have enough space to store all five pages and on stepping back from page 5, page 4 has to be recreated box by box as you step backwards.

All the optional boxes have labels that are self explanatory except perhaps the box labeled R( ). This stands for reliability factor, a technical term for a number measuring the quality of agreement between theory and experiment in a LEED structure determination. There are various types of R-factor and the table has a box next to R which is used to specify which type has been used, eg: ZJ for the Zanazzi- Jona version, P for the Pendry version.

6.3 COMPLETION OF THE MANDATORY BOXES

The program assumes that it is dealing with a periodic surface structure defined by a unit cell having the property that it generates the entire surface structure if repeated throughout the surface plane. A full definition of the unit cell can be found in the "International Tables for X-ray Crystallography" (Kynock Press: Birmingham). All that is necessary to define a surface structure is to specify its unit cell. This is done by specifying the lengths of the two sides of the cell, a & b, and the angle between them, A(a,b). The sense of A is defined by looking down on the surface from the vacuum, and measuring all positive angles in an anti-clockwise direction from a to b.

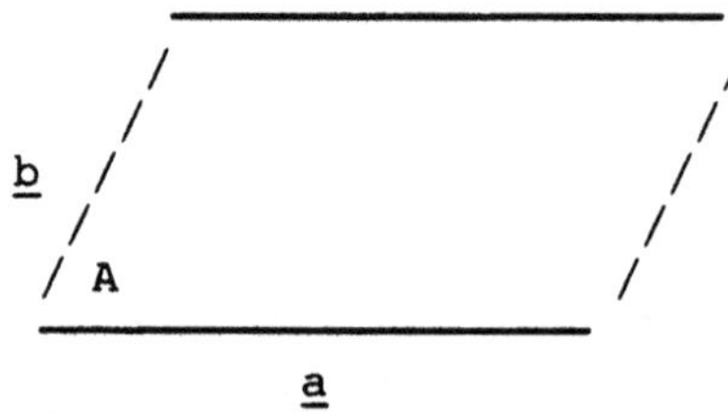

Having defined the unit cell the next step is to specify its contents. For this purpose the structure is divided into layers of atoms. Each layer contains one atom per unit cell, and is defined by the six boxes,
LAYER ATOM ATOM POSITIONS NORMAL LAYER SPACING ERROR
To define each layer it is first given a label: usually A1, A2, A3..... for adsorbate layers and S1, S2, S3..... for substrate layers. Next the atom type is entered: the chemical symbol being given in the form H, He, Li, etc. If this format is not used the program cannot recognise the atom type and will not be able to assign an atomic radius from its internal tables. <u>It is also important that the entries all begin with the first letter, not with a space.</u> Otherwise the program cannot recognise the atom type.

The position of the atom within the unit cell is specified in units of the cell sides. For example the atom shown by the asterisk in the cell below,

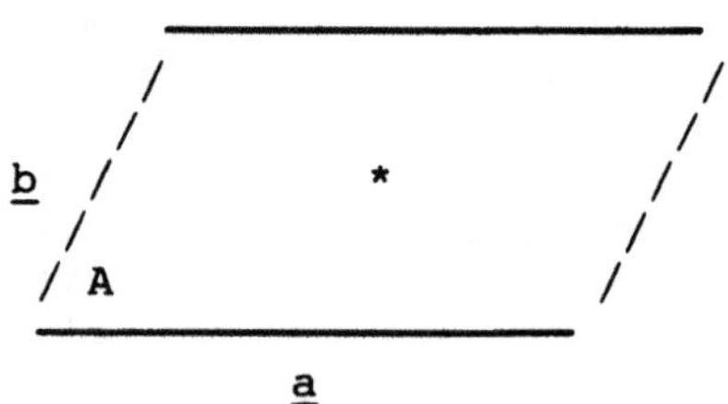

would have the following two entries under ATOM POSITIONS,

0.5 0.5

The box labeled NORMAL LAYER SPACING defines the distance between the current layer, and the next one deeper into the surface, the distance being given in Ångströms, or whatever was the choice of units for the cell sides. This distance may sometimes be 0.0 if there are several atoms in the unit cell at the same depth below the surface.

The final box gives the accuracy to which the normal layer spacing has been determined. It is not essential to the graphing routines and can be omitted.

<u>limitations on the number of layers</u>

As many layers as desired can be entered up to a maximum of 55. In the graphing routines it is usual to surround the unit cell with its eight nearest neighbours to give the impression of symmetry.

The entering of layers is terminated by a row of blank boxes which the program recognises as the end of this section and moves on to the next set of boxes. More precisely, the program looks for a space or a blank in the box corresponding to the second ATOM POSITION.

6.4 FILING THE STANDARD STRUCTURE FORM

After completing the COMMENTS box press ENTER until the subsidiary revision menu appears. The structure can now be tested as described in 4.4, or graphed as in section 5, and then revised if necessary, see 4.2.

When the structure form has been checked, it can be filed for future reference. Starting from the subsidiary revision menu, press F to select the subsidiary filing menu, followed by A to file the new structure. You are returned to the main editing and input menu. For subsequent access to the new structure you must return to the main utilities menu.

GPSR Compliance

The European Union's (EU) General Product Safety Regulation (GPSR) is a set of rules that requires consumer products to be safe and our obligations to ensure this.

If you have any concerns about our products, you can contact us on ProductSafety@springernature.com

In case Publisher is established outside the EU, the EU authorized representative is:

Springer Nature Customer Service Center GmbH
Europaplatz 3
69115 Heidelberg, Germany

Batch number: 10682317

Printed by Printforce, the Netherlands